UNFORESEEN UNIVERSE

UNFORESEEN UNIVERSE

DAVID GRASICH

ARCHWAY
PUBLISHING

Archway Publishing books may be ordered through booksellers or by contacting:

Archway Publishing
1663 Liberty Drive
Bloomington, IN 47403
www.archwaypublishing.com
844-669-3957

Interior Image Credit: david grasich

ISBN: 978-1-6657-3040-2 (sc)
ISBN: 978-1-6657-3039-6 (hc)
ISBN: 978-1-6657-3041-9 (e)

Library of Congress Control Number: 2022917340

Print information available on the last page.

Archway Publishing rev. date: 01/11/2023

CONTENTS

INTRODUCTION

Here is a guide to the unforeseen, thought up by an ordinary individual with an extraordinary imagination and new ideas. Newton and Einstein were both open-minded and self-taught master thinkers, obsessed with making sense out of reality. But what is reality? Who are we? How did the universe get started? Was there a big bang?
I would like to explore some possible answers to these questions.

I'd like to recount an unproven but new theory of what could have taken place in the history of the universe and how it relates to the everyday minds, choices, and issues of the world. Right beneath our noses in our expanding universe are weak and strong forces that remain hidden. Is this because it's right in front of us?

Science has never been able to explain where these two energies came from and how negative energy got started.
Only gravity can afford the power to create something new. Weak and strong forces are opposites, and opposites will always attract. This will force expansion and is the very beginning of equals and opposites that will become the new laws of relativity.

Dividing gravity is the only idea that can explain all the building blocks inside the expanding universe-combining to make a perfect circle. Understanding how energies are connected to each other under certain laws discovered by Isaac Newton has opened new windows, and one of these windows reveals a chain reaction that could have taken place that makes all energies equal. Each one of these energies has a positive or negative charge that is opposite to its companion while sharing the same laws. This could be the best way to comprehend how everything works: Relativity or reality are both the same, and where they came from could take us out of what I call the Jet Age.

In the upcoming text, we'll cover the following subjects:

The Five Rooms
Conscience
The Fifth Room
Implosion
Gravity
Constants (Chi)
Visible Light
The Force/ Negative Energy
Time
Dark Matter

THE FIVE ROOMS

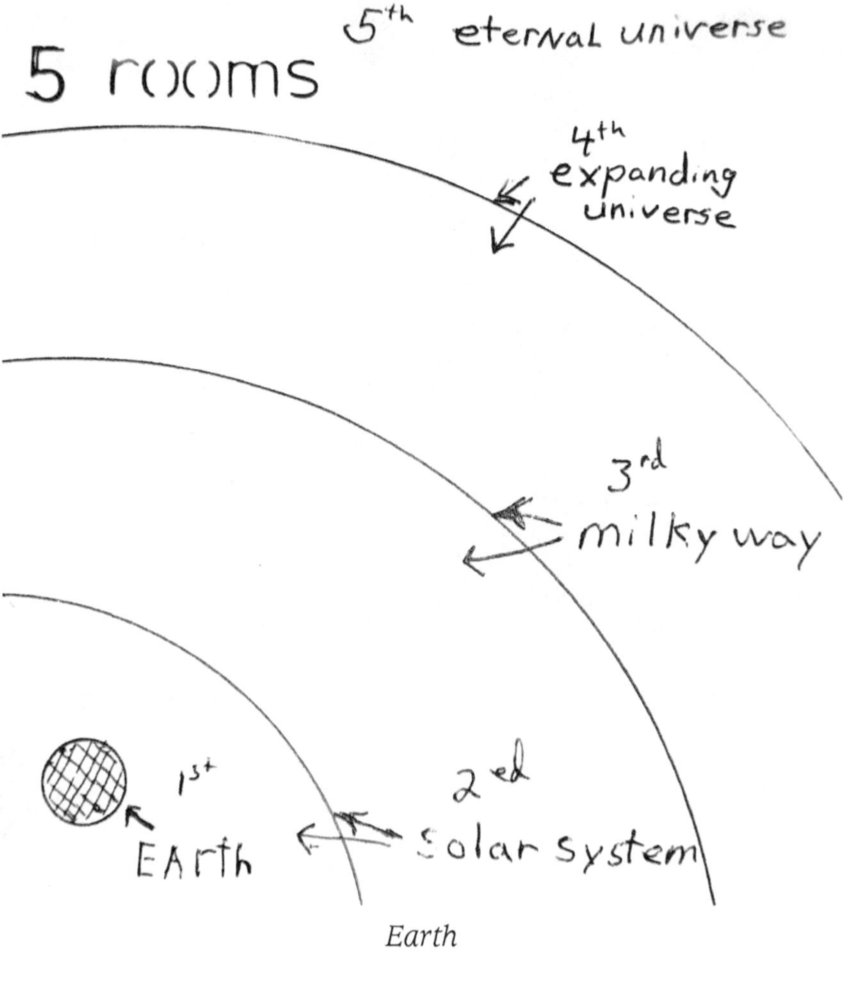

5 rooms

5th   eternal universe

4th
expanding
universe

3rd
milky way

1st
EArth

2ed
solar system

Earth

The five rooms represent (1) Earth, (2) our solar system, (3) the Milky Way, (4) the observable universe (space), and (5) the eternal universe.

Space is the observable universe as far as the Hubble telescope can see and what I call the fourth room. Could this fourth room be an expanding cold bubble inside an endless ocean of heat that is the eternal universe or the fifth room? This last room has no walls and is eternal, but inside is a massive bubble called the observable universe, which could have hundreds of billions of galaxies. Somewhere in this galaxy belt is the Milky Way galaxy, the third room, which is a hurricane of spinning stars that will contain our solar system (the second room) and finally Earth (the first room).

Earth is most certainly spinning in three different directions. The observable universe is surrounded by the fifth room. The fifth room is eternal and has no walls. In this place with no beginning or end there is an eternal pull called gravity. Regular gravity is the most powerful force that exists. Inside this powerful Force is a bubble called the observable universe. This fourth room will have a weak gravity called the weak force. Here was a powerful energy that was reduced to create something new-like negative energy, this power had to come from somewhere and could only have come from gravity. This is the point at which gravity could have been divided to start a new universe. In order to divide a universe you have to divide its power.

Dividing gravity in half will leave behind a weak gravity in the fourth room. Dividing something like gravity meant that these two energies divided and split the universe in two halves, bubbling out from inside in the form of a orb. This is the fourth room that follows the big bang. The weak force and the strong force could be the secret that creates all circles in the new universe. This orb can only expand or collapse. All energies from the big bang were divided evenly with an opposite companion. Weak gravity was divided from regular gravity. Negative energy was divided from positive energy. These are all opposites.. however in the fifth room, gravity is the only force that can expand the fourth room because regular gravity in the fifth room will pull weaker gravity in the fourth room and force expansion. The slowing down of gravity will force the universe to expand.

Gravity has been divided in half in this new expanding universe. Gravity will also slow down time and create a bubble or a black hole. The strong force is pulling the weak force.

Gravity and the fourth and fifth rooms

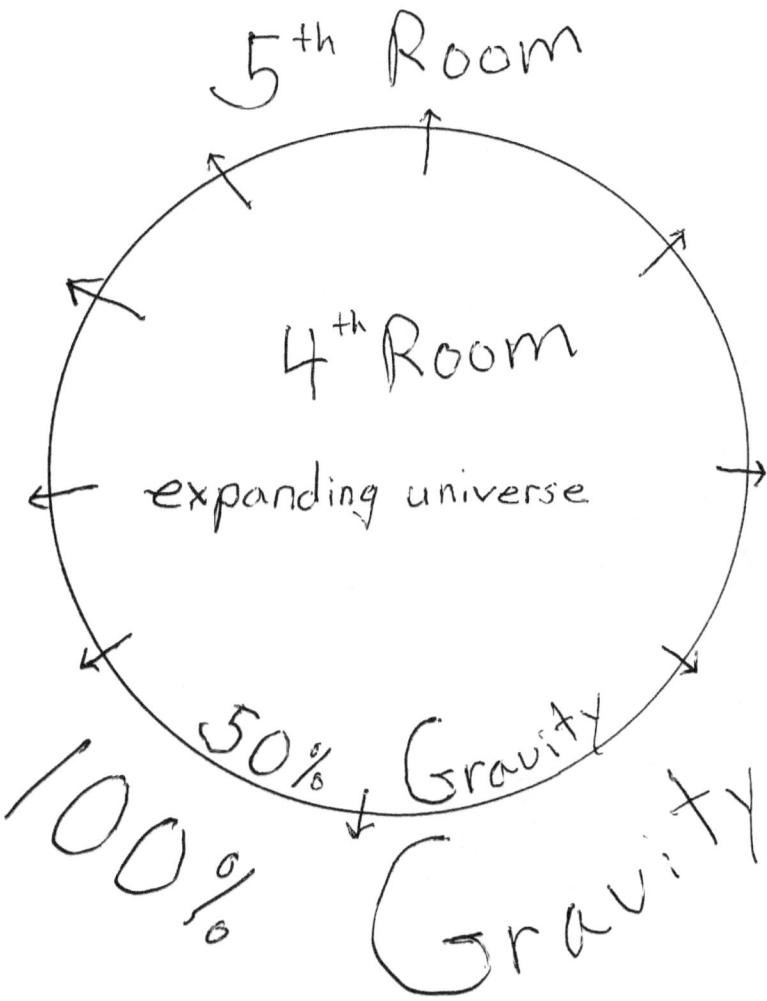

Gravity becomes stronger after leaving the fourth room and going to the fifth. As speed picks up, so will time. However, the fourth room is a growing black hole that nothing can escape. It's being pulled outward in every direction by the fifth room. Only gravity has the power, called the strong force, to do this. This force is pulling the weak force into the form of an expanding orb.

Fifth Room
Eternal Universe

Gravity
Strong Force

What is certain is that the fourth and fifth rooms are vastly different. In fact, they seem almost like opposites. Space is just about completely void of energy. Negative energy is simply a complete absolute equal opposite of positive energy. This is very important to understand because all major energies in space will have an opposite companion that is separated by charge, either a negative or a positive charge. Space is a frozen bubble surrounded by an eternal pull of heat. Hot and cold, positive and negative—these two sets are the same. Two energies divided at the beginning of the new universe, equal in power but opposite in function but share the same laws. There is a reason for this.

Isaac Newton was the first to discover this new understanding of relativity and our reality. In the beginning of the new universe there was a chain reaction called equals and opposites. This created new sets of laws from the very beginning until now.

Matter or mass itself has an opposite called no-mass, but what is no-mass? The answer is very simple: It's called space! Everything has an opposite, even words or emotions that have been divided from good energy to bad.

Every good word or emotion will have an opposite. Have you ever wondered why? Everything is divided! This ultimately means that we have been given conscience. Conscience is a reflections which means we see choices, like pushes or pulls, or good and bad emotions. Trying to comprehend without negative and positive energy would mean our minds could not balance between choices. The laws of equals and opposites creates balance and balance means choice- but what is choice? Ask your conscience. It will say you're now alive, and life comes with choices! Every day will come with lots of simple choices and issues to deal with, and some won't be so easy to make. Choices can have consequences. Good choices can be dull and hard to notice, but submitting to bad choices will always somehow find you. Life can be a slow walk to the end.

CONSCIENCE

Conscience is that faint light that illuminates a hard-to-see yellow-brick road. Think of your conscience as an iron ball in the center of a long, flat magnet. So there you are, alive in the center between two energies (positive and negative) that dictates everything. Each will have a slight pull. Rolling too much to one side will narrow your view from the other. If you can balance your conscience, learning to think ahead can be easier.

Always be prepared to expect the unforeseen. Thinking is a process we learned mostly from others, but from ourselves as well, in order to mold together a conscience. People who hold hands and follow strangers cannot see their path. These are good people who lack insight and personal freedom. Some of us can stray like sheep, we get caught grazing forward, without realizing where we are going.

Conscientiousness means you can have your own choice, like a quiet voice that shows two sides or a shadow drifting between loud voices. How do you respond to these voices? I could only suggest to smile and be courteous, like giving a big tip to a bad waiter. Be aware that your choices can have a mirror effect on other people. Tomorrow is a new day, and some people can change. This conscientiousness can be a very powerful force and could flip backward when a relationship goes stale. The one who is being pulled into another relationship can become a bad waiter, but what about you? How would you handle this if you got dumped by someone you love? What's your conscience saying?

Step back and be aware of these overwhelming emotions from anxiety to extreme jealousy that are being hurled at you. These conflicts will only add to more confusion. While eyeballs are gazing, anger will blind your path. Try being mindful of this conflict. Love can cloak itself to look like hate, but these are opposites and opposites can easily be flipped backwards. Consciousness is a choice, choices can become compressed. A good conscience is like a compass. Let it show you the way. Life is a never-ending battle between people. If you can keep to your commandments, you could find contentment. Your conscience will teeter with choices. Try to always balance your emotions and follow your good nature because most bad choices will come to you. Bad choices can become clouded, revealing this future will be difficult, and only good can unveil one's true nature. Try to discover this other side of yourself, and remember that this conscience is always looking. Can it see what you see? Who is in control (me or I)? Seize this opportunity called life that's been given to you. Could it be possible that everyone's candle was created from positive and negative energies with something else—perhaps a divine light, a conscious eye that never closes, or like a faint glow from a lighthouse, as you drift in an ocean of waves? Could this beacon be a warning sign that sin could cloak your vision? And just like an iron ball, you could roll off your path like on a magnet. The farther you roll, the harder it pulls.

Bad decisions will complicate your choices. A real conscience or a good friend with a clean nature can help center your choices back to a normal direction. Use the force; it's always there. A true Jedi can see with two eyes because one force or eye (+/+) has been flipped backward. Conscience itself is like looking at two mirrors, but one mirror has been flipped backward to a negative mirror (+/-). This will reveal a third mirror, and that is you! Hear is you looking at two mirrors. This is a pyramid effect for continuous reflection. If you flip the negative mirror back to the positive side, you will disappear with no reflection. Your natural ability to chose will be gone. Flipping one energy to its opposite creates choice, and choice is a balance between two energies (+/-).

Everyone exists in a world of good and bad, and from birth to the now is a balancing act of choices. Soon our paths will end, and those choices will follow us to a scale,we could be balanced.

THE FIFTH ROOM
(ETERNAL UNIVERSE)

This eternal room is like an ocean that surrounds a sea that I call the fourth room.

Before the big bang, the old universe was a dead place—no light or electricity. We don't even know if this place had heat, but if it did, this heat would have been a by-product from gravity and mass creating a heat. It is believed that there were only three energies that existed in this place: heat, gravity, and matter. Gravity is at full strength throughout the eternal universe (the fifth room), but we know in our new universe (the fourth room) these laws of relativity could be different. This means that gravity must be different?

If it is possible to travel beyond the expanding universe to this fifth room called the eternal universe, could such a universe actually exist? What would be there? Most astronomers will agree that this place exists with only one word to describe it: chaotic.

Just think of this fifth room like a never-ending puzzle game. Each puzzle piece is separated. This room could be set on a high temperature. This heat could keep everything separated, In between this eternal separation is a grid or a web of connected and compressed energy called gravity. Gravity is like a high-powered wire wrapped around each puzzle piece. Every puzzle piece could have hydrogen that's under constant stress from gravity. This gravity is a strong force, and in this room this force is powerful and never-ending; this gravity force is the power, and raw hydrogen is the fuel that's expanding the fourth room.

These fourth and fifth rooms could be a never-ending tug-of-war, but the fifth room will always be winning, while never getting any smaller compared to the fourth room. There are only three words that can explain this place: chaotic, never-ending, and incomprehensible.

IMPLOSION

What could the implosion before the big bang have been? Implosion could have happened so fast that it could have actually taken infinitely no time to collapse, and no time has an opposite called infinite time. Is this a real question? What is real is that the implosion is the only event we know about that occurred before the big bang. Here is an equal and opposite reaction. Implosion and explosions are opposites,but equal in power.

Newton's laws of motion shows you how an implosion can be equally as powerful as the big bang. Newton was a master Jedi, and his law states that every action will have an equal and opposite reaction. So is the big bang a reaction from an action called implosion? And did the singularity blow up? These are questions that have never been answered. The singularity did not blow up! With no metals, the singularity simply broke up evenly following the big bang, and everything started from there. The singularity is believed to be a massive ball of mass before the big bang.

This singularity was pulled apart from the vacuum pressure of this new universe. Everything in space expands and will follow a never-ending explosion. This means the entire fifth room is blowing up, but this is the only way expansion can happen...

Now mass has an opposite called no-mass, but this no-mass would have to be negatively charged to become a complete opposite of mass. Space just happens to be negatively charged and there's nothing there. A real definition for space would have to be negative energy without mass or the exhaust from a never ending bomb. Mass and space are absolute opposites; these two laws were set perfectly. This is an opposite that hides in plain sight. Most of the fourth room is empty space with almost no resistance.

Could the implosion itself have something to do with the expansion rate taking place now. And if so- was this a recoil effect, which could have a limit?

After the Big Bang - fifth room

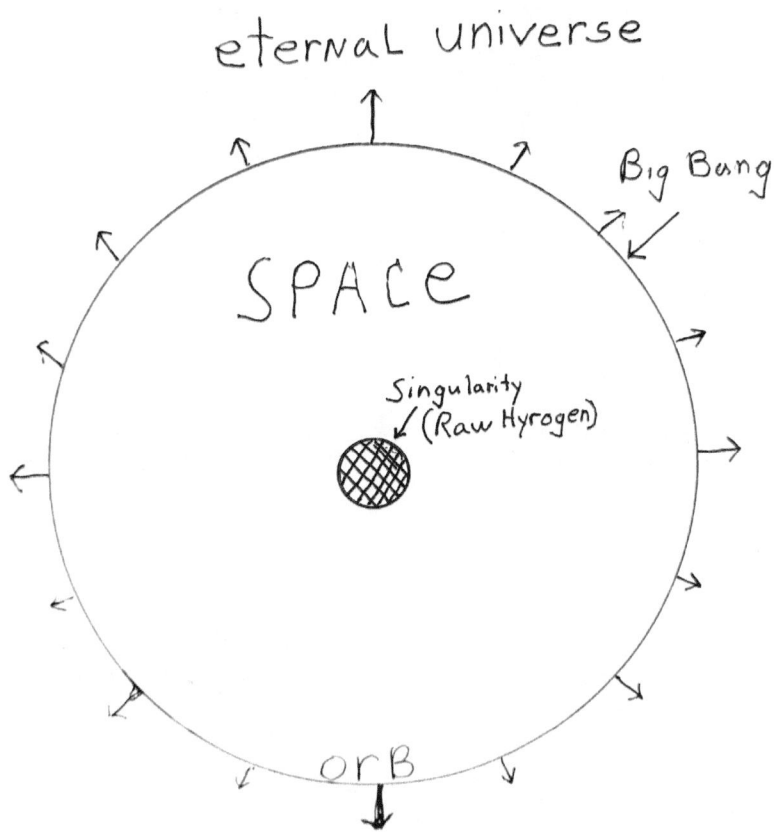

GRAVITY

25

This force is unlike anything else. Its energy source can only be multiplied by infinity. This fifth room, that has no end or beginning, will have an eternal grid of connected and compressed energy called gravity, which is a pulling force. Gravity is a force we feel everyday and every moment - but it is only when we sleep that this force is conscientiously voided. Gravity is the Ultimate energy, this force dominates and controls everything in each room. This includes the first room Earth.

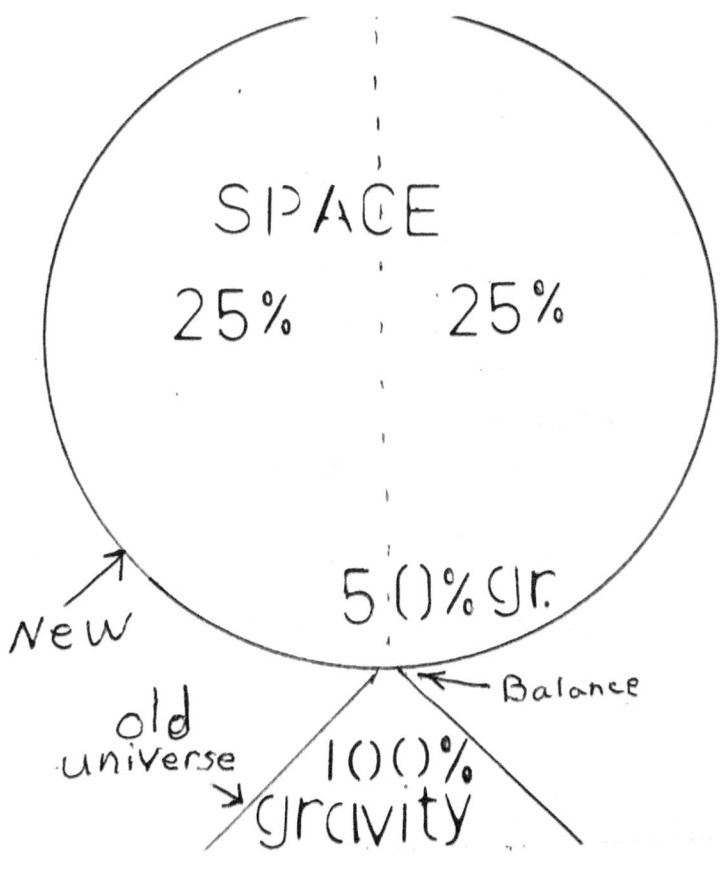

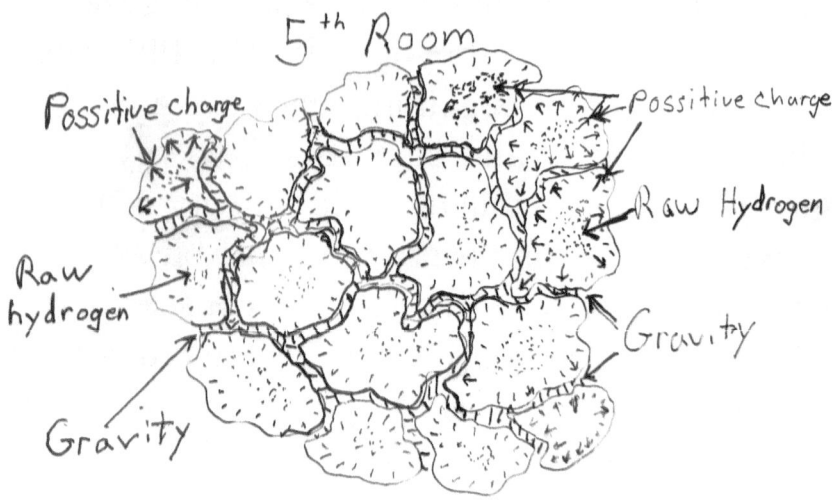

Over a century ago the idea of using crude oil to counteract gravity's pull sounded like a great idea, but it went terribly wrong.

Now oil has made everything easier. Except it's getting harder to breathe.

This new source of energy did help stop the wailing ships. But it has resulted in one of the core reasons for overpopulation of humans and will help to deplete a very important food source (reefs). This has lead to a bleaching effect with the Earth's many oceans and is one of the reasons I chained myself to a tree, because I can't breathe underwater to save the coral.

Gravity is still the same. Weak gravity is a force dividing outward in every direction. Strong gravity is pulling weak gravity and a never-ending explosion toward it, turning heat to cold. This cold charge called negative energy is the exhaust from this never ending bomb (the big bang) and will fall into the skylights of the space bubble, compacting and forcing expansion. As negative energy fills the space bubble with no energy or opposite to the fifth room, opposites attract. The fourth and fifth rooms could be eternal, a never ending push and pull. Space is always growing but will remain the same, always growing but never getting bigger. Compared to the fifth room it is overlapping, time will bend opposite, or backward, to a perfect curve called an orb (the fourth room).

Try to comprehend what's really happening, and then—and only then—it will make sense. The old universe has collapsed in, and the new universe is expanding out. This is the opposite reaction. Opposites divide everything. .

This Last room is being devoured, burning up its positive energy and falling into an empty void called space, as a complete opposite.

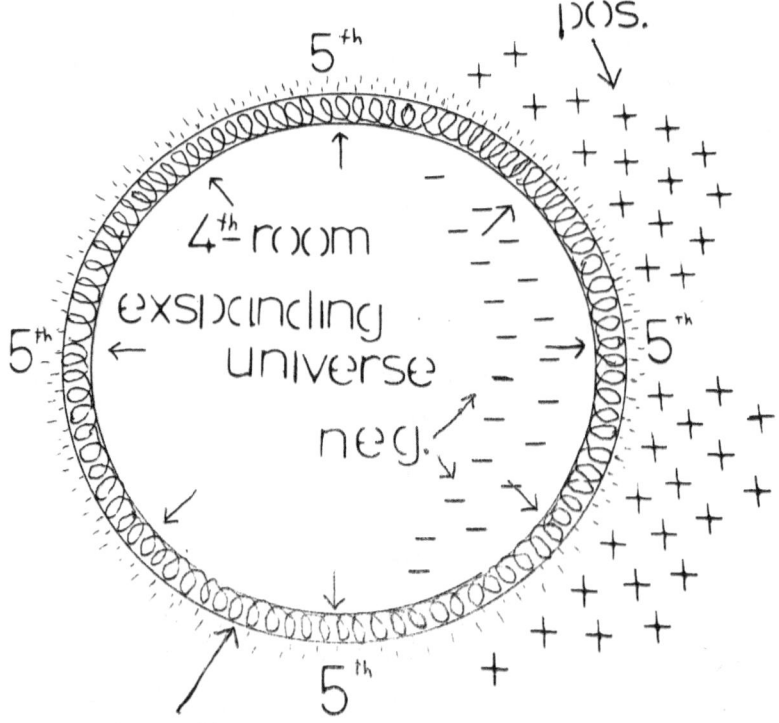

ETERNAL UNIVERSE

pos.

5ᵗʰ

4ᵗʰ room

exspanding universe

neg.

5ᵗʰ

5ᵗʰ

5ᵗʰ

opposite temperature flip
(neb)

CONSTANTS (CHI)

A circle and a hexagon are two symbols sketched throughout the cosmos. A circle is a visual representation of how strong gravity and weak gravity are combined. A hexagon is a weak but stable force called constants. A hexagon is a visual representation of how all six building blocks were forged from the big bang and have become separated evenly with an opposite partner. NASA's Voyager 2 confirmed that a hexagon can be as large as Earth as seen on Saturn or as small as a snowflake. A hexagon has six lines representing six energies. Each energy was split apart from its equal to become its opposite. Negative energy comes from positive energy. Weak gravity was divided from regular gravity, and space was divided from mass. All share equal power—one to the power of six.

This is why snowflakes are framed with six sides. Science has always wondered what forms a snowflake. It's like asking, Who built this house? Well, find the framers and answer the question!

Constants are the framers and will always form a hexagon when united with hydrogen under the right conditions. Just look again at the sun behind a lens, either on TV or in pictures, and notice that most of the time it will form a hexagon with six coronas. Constants can be a real freak of nature. It's like a compass that points everywhere. The planet Saturn, snowflakes, and the sun all share two things in common. First, they all form hexagons, and second, they are mostly made up of hydrogen. The new universe is packed with this type of matter, and it's because the old universe has only one source of matter. This type of matter has to be raw hydrogen.

This would mean the singularity could have been the largest ocean ever conceived before the big bang. This massive ocean would have been a positive charge, like that of a car battery. Creating negative energy would have been impossible without this ocean. This ocean could have been sucked apart by the vacuum pressure from this never-ending bomb.

If we could just pull out of the Jet Age, this would be a big leap for humanity—but a dangerous one. However, remaining dependence on oil and coal could be just as bad. All six energies were created equally at time zero; a perfect circle is formed by two energies: weak gravity and strong gravity, but constants form a rigid circle. Each ridge represents a different energy but units to a circle (chi).

Cold, or space, did not exist before the big bang. Space was created with everything else. Space is a perfect light bulb. In space negative energy is like extremely dry paper, and positive energy is the water.

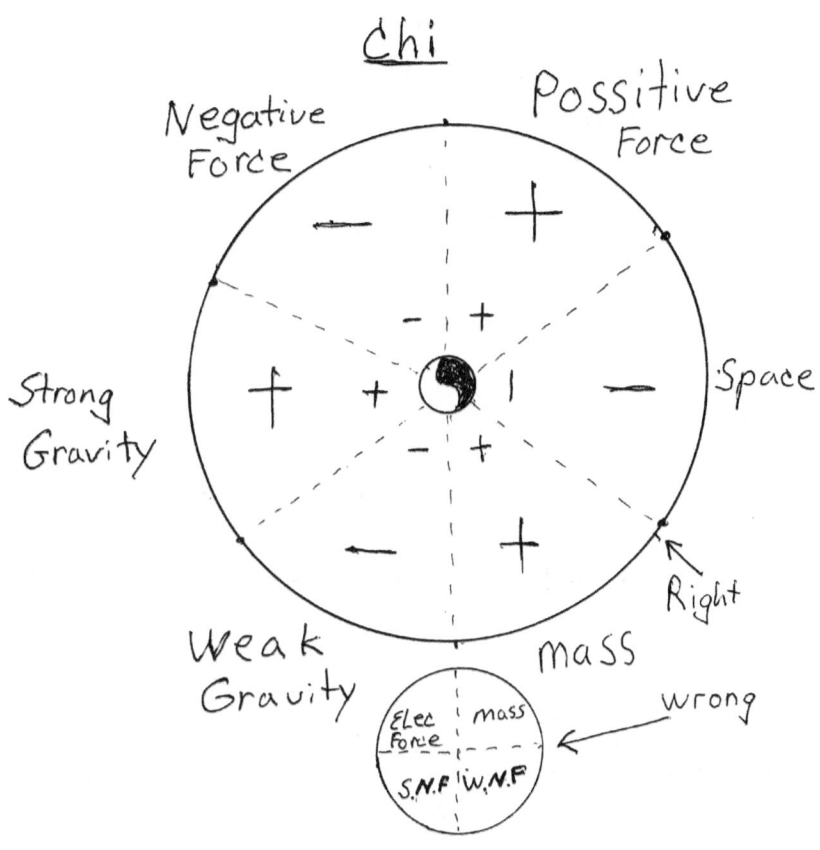

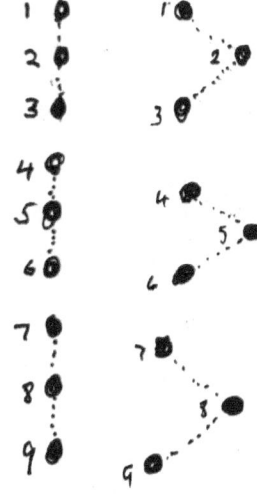

In the New universe we are only Given Nine NumBers. Light will Have to Obey these Laws of math. These Laws could Form a Crystal eFFect which could ReFlect on each Side. Multplying three should Fill all voids in Space

When they get too close, they will absorb each other evenly.

Space will absorb light energy and multiply it. Negative energy will push and pull light to its farthest limits. Light is like the center of a flower, surrounded by six energies. Light is like radiation being sucked out of the center of six energies called constants.

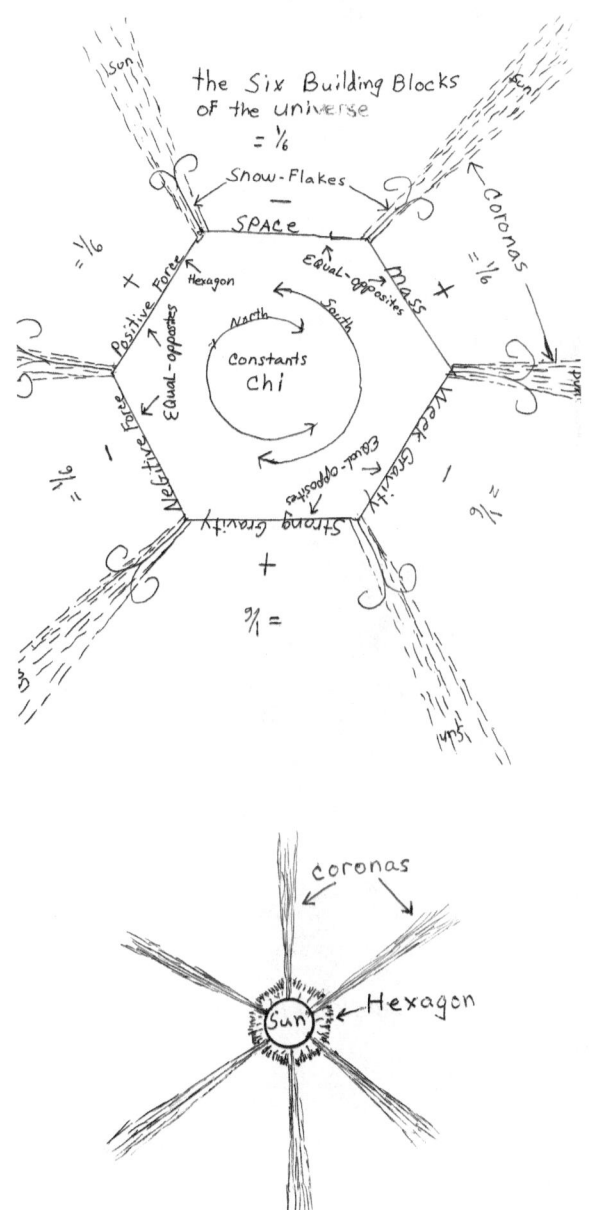

One light crystal will multiply by two forever by negative energy until it runs out of positive energy.

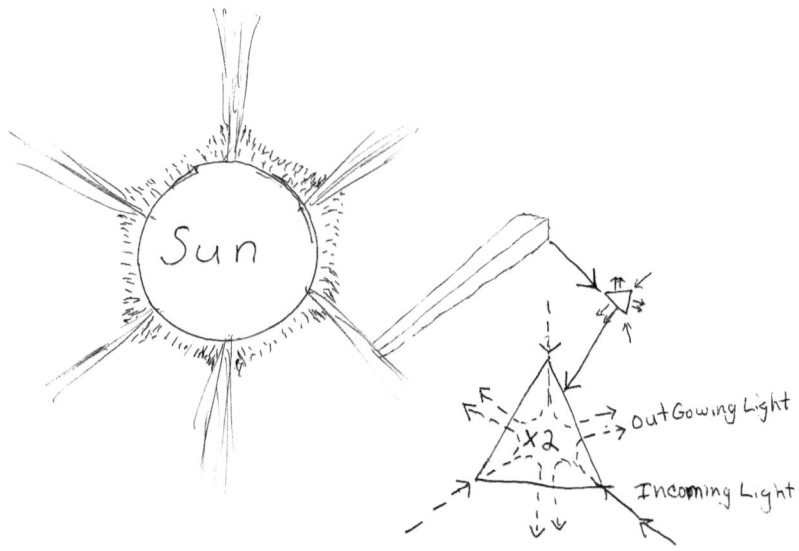

A three-sided light crystal

VISIBLE LIGHT

To truly understand visible light, close your eyes. Visible light is the purest, safest, and cleanest energy ever created. Visible light will simply reflect off itself like a secretion of two pure, evenly separated positive and negative energies that are opposite in function but equal in power. These two energies will come together and radiate out many different spectrums, but the very best is visible light—a white rose that shines completely and in color. Light is too fast to see; however, it could be possible that visible light could take the form of three-sided light crystals. This would be stretched and multiplied by negative energy.

Light crystals are constantly reflecting off themselves and radiating out more light that is getting divided by negative energy. Negative energy is the energy source and the secret for spreading out more light. Could three-sided light crystals have a mathematical advantage, creating perfect efficiency with no voids? But why number three? In math numbers only go from one to nine, and ten is just a reload back to one. Three is perfect. It can form a pyramid effect, a three-sided light crystal. Could light simply be math? It's light math, and three is the only clue. Just look at the sun through a camera's eye or on TV and take note—the sun will reflect six coronas most of the time. No big deal, right? But look again! Science refers to this as a double-lens effect.

I see a conflict of interest.

If you take an even number, six, and divide it in half, then you have an odd number, three. However, coronas only come in pairs, and three cannot divide but will get multiplied and spread out. $3 \times 3 = 9$, and ten is a reload back to one. This means three will take up three numbers to nine. It's perfect for light!

Light crystals are being stretched and multiplied by negative energy, and without the vacuum of space all forms of light would have nowhere to go. Space is negative energy. It's like a liquid mirror effect that stretches from one end of space to the other. Light crystals are bouncing everywhere. Cold energy is the opposite of hot energy. Opposites attract, and this is the secret of visible light. Visible light fits right in between these two energies and becomes a clean, safe radiation that is united in an eternal marriage by weak gravity. Try thinking of visible light as a flat mirror folded on three sides, a pyramid shape that is stretched out like a Pinocchio nose pulled by negative energy for billions of light-years long. Space has no resistance for visible light; it's being pulled everywhere with no voids.

In space or on Earth positive and negative come together like hot and cold.Gravity will unite these two, creating a clean radiation called visible light.

From our sun, visible light is created from heat, amplified by sound, and sucked out into space by negative energy. All this will come together with almost no resistance.

Look at our sun. It has the loudest trumpets in the solar system, and all that noise is absorbed by the corona and negative energy; however, you can see it. Visible light is the most incredible energy ever created, and until we die, light and life are connected in ways we don't understand. Light can show you the way or blind you. Try stepping into the shadows. Turn and gaze at what's been previously glaring. Toss your horse blinders, and always be mindful of crab traps made by man for the visually impaired—us.

Visible light is being stretched and turned like a helix. This light shines complete, exposing everything. Nothing can hide from light and its creator. Without light everything is lost. From this light we come, and then back we go.

We live out our lives like a candle lit by our parents; don't blow it!

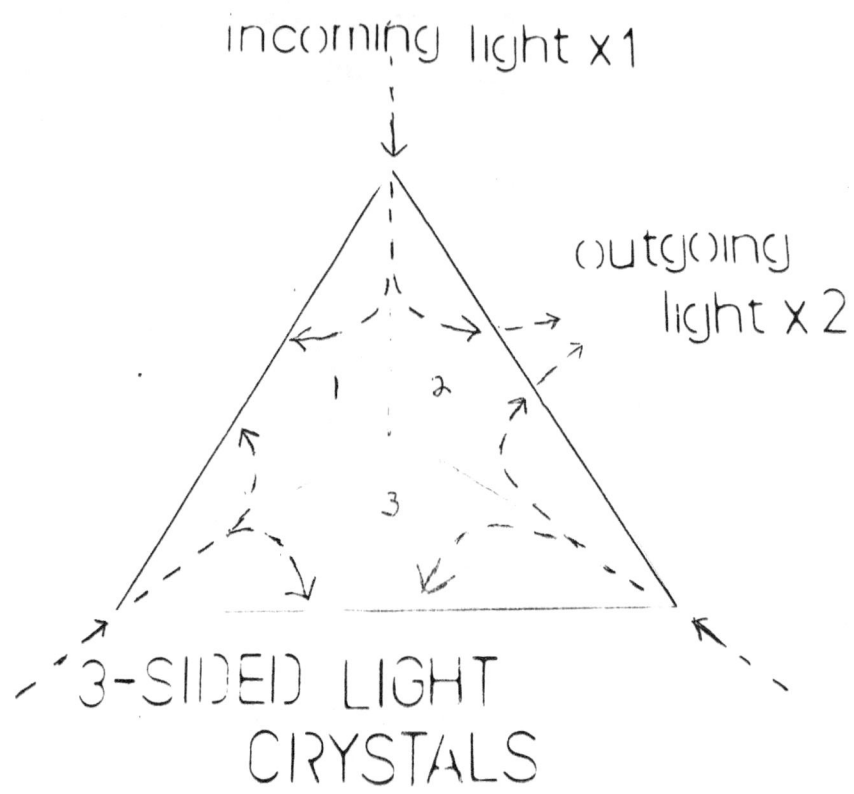

incoming light x1

outgoing light x 2

3-SIDED LIGHT CRYSTALS

THE FORCE/
NEGATIVE ENERGY

Lucas is right, the force is all around; it's just that we don't believe! That's us, and what's really fascinating is that it's true. Negative energy is the glue for everything. Cold energy is like a magnetic glue, binding everything together. Cold is a pulling force, and heat is a pushing force. Cold energy is the exact opposite of hot energy. Cold and heat have to obey three laws: Direction, temperature, and charge. (1) Direction (down, up)—down means that this force points toward the weak force, and up means the strong force; (this is the eternal universe), (2) charge (north, south), and (3) temperature (hot, cold). This is a very unusual triple-opposite. They're just like three mirrors reflecting only on themselves.

This is no coincidence. Creating energies will force laws to serve a new relativity, and a chain reaction will follow. Everything in nature and in space will show you clues that are right in front of us; that's why we don't see it. One of these clues is a freak of nature called division. Everything's divided. Just like polarity is north and south, this is just another opposite that makes up the electromagnetic- glue. This glue force is all around us. If we teeter this force with balance, common sense could take us farther .. Generally speaking, If making money always comes first, and family is second. I would suggest- jeti your conscience to a new level. Mother Earth will give us clean air to breathe in, but what do we breathe out? Unfortunately, polluting the air with words can become easy. This can reflect off of weak minds-like a broken mirror with no answers, just excuses. Never let your guard down, or you will get sucker punched. We believe we are awake, but are we half asleep? Close your eyes, look around, and awaken. A real Jedi can calm his or her emotions, visualize between two forces (+/-), and gaze at reality; while achieving contentment through the letting go of selfish desires. exit your surroundings from concrete while indulging your visual consciousness with nature. The star lights at night should be a constant reminder of our curious nature of the unforeseen.

These two forces, positive and negative, have flipped everything half backward and only make sense on a magnet or in a mirror. The electromagnetic is what you get in between and the glue that binds everything together. The creation of negative energy was the beginning of everything new. This is the opposite energy from heat.

If you look at a mirror, you'll only see half of you. Move your right hand, and the mirror shows you left. A mirror reveals a half flipped reality. It's simple, a mirror will show you an opposite direction This is how everything is set.

If you put two opposing sides together, you have opposites, just like a magnet. Black and white, hot and cold, up and down, day and night—this phenomenon could go on and on. These are just simple examples of these freaks of nature that sit right beneath our noses, hiding in plain sight. Everything's divided. If you take white and flip it to its opposite, then you'll have black. Now look again. If you take black plus black, you'll have two equals, but if you flip one equal backward you'll have equal opposites. Without equals you can't make opposites. Opposites are the foundation of creation- like a mountain that can't move. Here's a good question to ask: If you take two bowling balls into deep space that are the same weight and size, and hit them against each other at the same speed, what exactly would happen? The answer is they will repel with equal speed in opposite directions, but is the speed the same, or will the speed divide in half? Fifth Room

5th Room

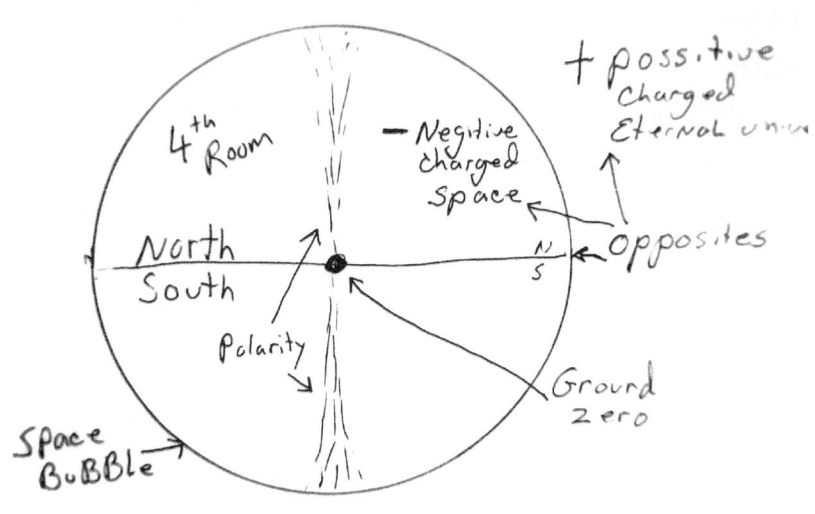

+ possitive
Charged
Eternal unien

— Negitive
charged
Space

North
South

Opposites

4th Room

Polarity

Ground zero

Space BuBBle

5th Room

EQuals

| |

+ +

opposites

1 2

+ —

Equals and opposites seem to play a big part in reality. The number two is linked to both of these words. If you take the number two and cut it in half, you will have an equal 1+1. So if you take one and double it, now it's the opposite. This shows you everything is divided, even math.

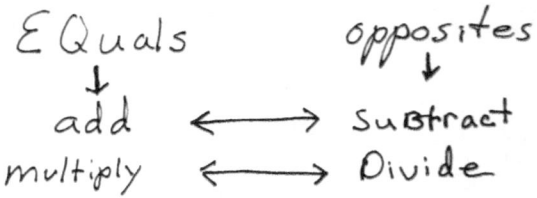

Equals and opposites seem to play a role with reality. The number 2 plays a big part with these two words.

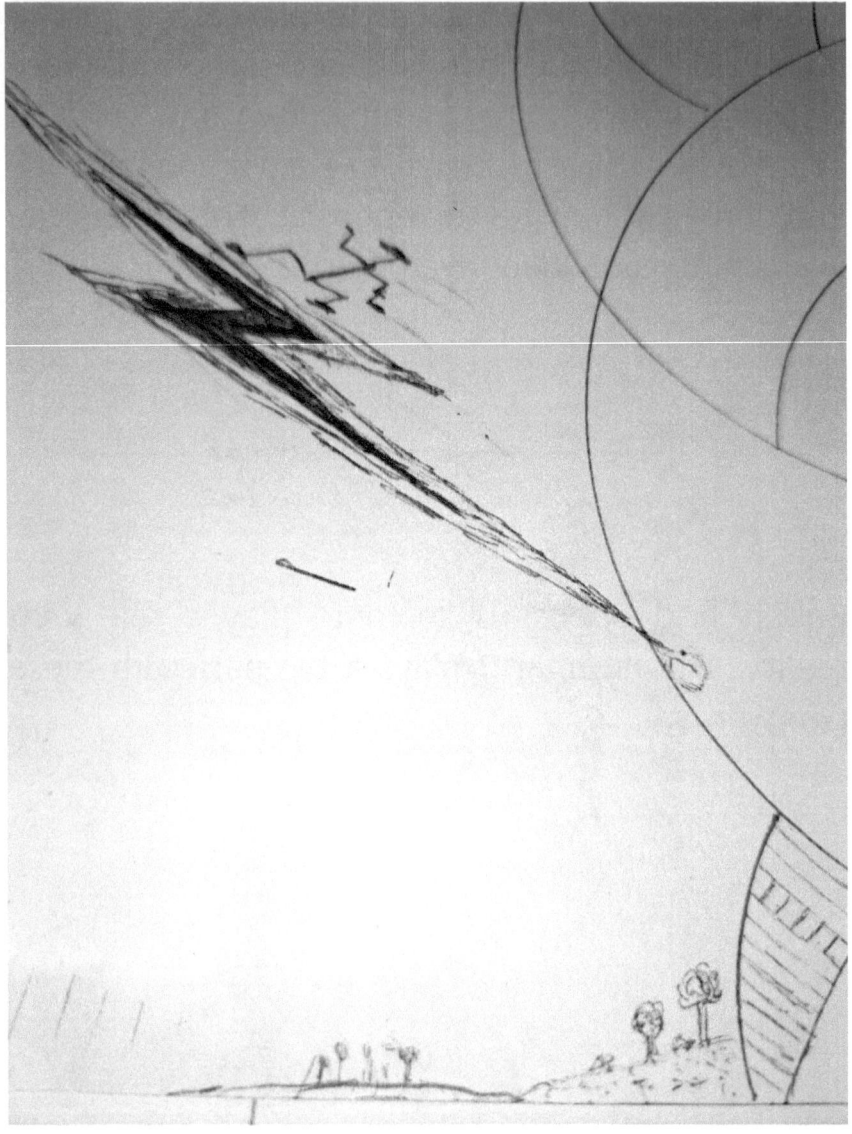

If you take this number and divide it you will have an equal, 1=1. However, if you multiplied 1+1, you have an opposite.

Dividing and multiplying are opposites but equal.

GEARS OF TIME

From Earth it's about sixty-two miles to the second room (solar system). From here it's over 12 billion miles to the edge of our solar system. Leaving the solar system to the third room (Milky Way) means Earth is spinning in three different directions. Could there be a fourth? Could the entire fourth room be turning? Going from room to room is a massive leap in distance, but this seems to have a conflict of interest in math. That might surprise you. Add it up. Is there a pattern?

Sixty-two miles (first room)

12,161,300,000 miles
(second room)

5,879x 1000000000000

(third room)
Third—Milky Way

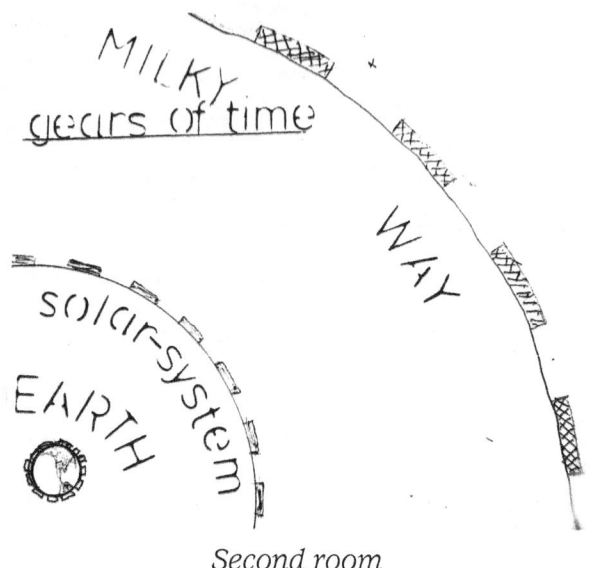

Second room

We still don't know the true size of the last arena or how fast it's expanding.

If you backtrack from these calculations, it should reveal a pattern. Each is like a gear. The bigger it is, the slower it turns.

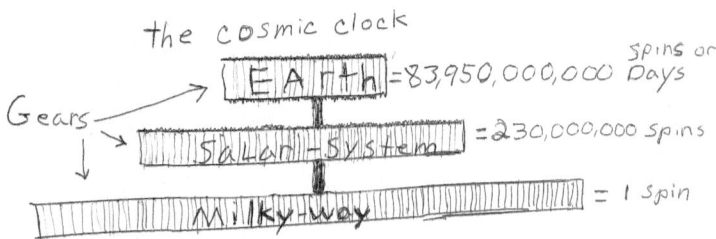

Earth=365 spins
Solar System=1 spin
Milky Way 0.0000036016= spin.

It takes 365 days for just one turn of the solar system gear. This solar system gear will have to spin 230 million turns to complete one turn of the Milky Way gear! Could there be a fourth gear spinning? And if it's turning, then how many times has it turned?

The graph above is similar to a clock.

Earth, the solar system, and the Milky Way are controlled spinning matter, divided by space. Space is divided from mass, only time can calculate distance.

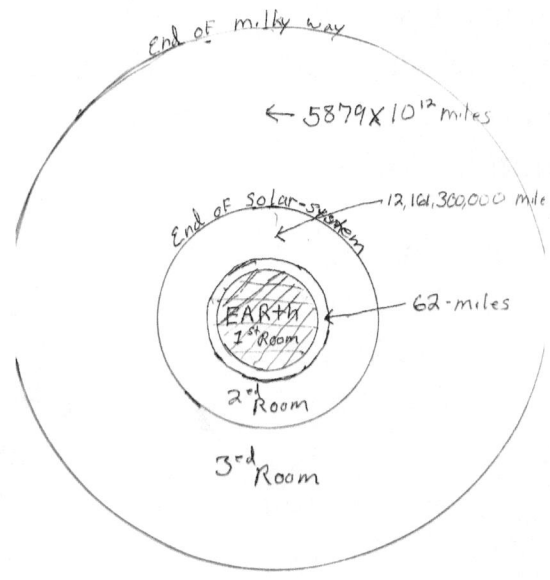

62 miles (first realm.)
12,161,300,000 miles (second realm)
5.879x1000000000000 miles (third realm)
The fourth realm is unknown.

What has the power to move mountains or map space? Only time can tell. However, time remained an enigma from the past until zero was conceived. Here was a massive breakthrough in math and time. Zero is a circle and a perfect symbol for time, so to really understand what time represents just look inside an old clock and take note. It's full of circles! Flip the clock around, and take notice of the second hand. This hand moves the fastest— here is Earth. The minute hand represents the solar system, and then the Milky Way is the slowest hand in this comic clock that represents the hour hand. From this view, step back and look at our galaxy. This is a colossal ballroom with almost a hundred billion spinning clocks.

Time must be round, but what makes a perfect circle? Could this be like a bubbling-out effect from two forces called weak and strong? Could it be a mirror image of the other half? Could it be a mimicking effect from the big bang? Just think of time as an orb. To draw a line around an orb to find the end means you'll never reach the end; it's infinite, just like time. Positive force plus strong gravity divided by negative force plus weak gravity equals a circle (time). The power of time has been given to us, but time will roll on and cannot be stopped. With time comes choices, and these choices are yours. Choose wisely—time is slowly running out.

One day in the fourth and fifth rooms

Time scale in the fifth room

Time scale in space

Space-time has been slowed in half, curving out like a gear, but outside the fourth room, into this fifth room everything is fast and chaotic. This is because everything is on a regular time schedule, but the space bubble has been squeezed into a circle (orb). Gravity has dictated time. In the fourth and fifth rooms, time is equal but opposite.

24 = 48, equal in time but opposite in function are distance. This means the fourth and fifth rooms are perfectly opposite. This should be an oohh moment, understanding what time means- is like understanding our universe and all of its gears that slowly turn. Equals, opposites, actions and reactions are the gears that turn the fourth room. Inside these four gears are the six building blocks. That's ten spinning gears. A clock is basically just a device that measures time in the form of distance. For example we use the term "light years" to measure distance between Stars. In the fifth room the distance is twice as far, and time is twice as long.

GROUND ZERO

61

Ground zero is the exact location where the big bang took place, and from there the observable universe expanded out in all directions. But where did the big bang take place? Could such a place even exist? Just getting to ground zero could probably only be done through thought or wormholes.

The idea of a ground zero would have taken place during and after the big bang. After the big bang, smaller galaxies could have slingshotted around larger ones to reverse course, and these new, young galaxies would have clustered up with others around ground zero like a beehive. Gravity has a way of attracting mass. In time these galaxies will beehive around ground zero and would have been the center of the space bubble. This beehive continued to reform to a more stable galaxy-sized black hole. This monster is the biggest of them all and would be crowned with a halo of massive quasore's and capped with two spinning witches hats, called polarity or north and south.

Traveling to a place like this would mean leaving our galaxy belt or the ocean of galaxies into the underverse. This is a very dark place where light is only coming from behind. It would feel like descending into a deep, dark ocean at unimaginable speeds to the center of the space bubble. If you stay on course, in time, you will start to see a faint light straight ahead. Soon the heart of the fourth room will be revealed. What will you find? How can you slow down? Is everything spinning? Does the polarity reach the top? And the most important question of them all: Is the space bubble turning? This would mean we exist in four spinning realms, surrounded by a never-ending room. Could the center of this fourth room have a black star as big as the Milky Way? Questions like this may never get answered, or even asked, but it's interesting to note that anything's possible in this new universe. Even today's most high-speed astronomy is based on theories and ideas that should lead to answers not questions.

We just don't know and may never know, but these ideas are based on curiosity.

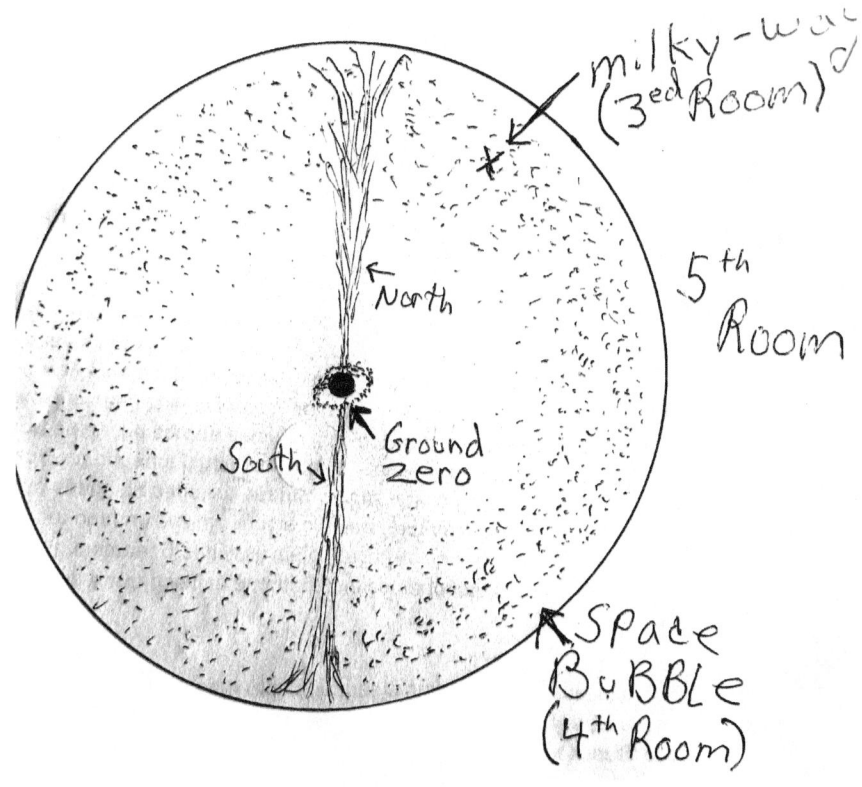

North or South
Milky Way

BLACK STARS

Every black star will have an event horizon. The event horizon is a massive supercollider that turns everything it touches into its opposite. Going from mass to no mass, it leaves only unknown radiation behind; however, the larger black stars have jets. Jets are holes that are coming from the north and south poles like on earth and are funneling escaping matter into new possibilities! This ejected material works like spreading out silt in a pond. This type of matter, or silt, is what creates new elements in deep space.

It's widely believed that most black stars come from massive blue stars collapsing or when two neutron stars collide. Another theory is when too much gas compressors from a nebula and forms a black star. When black stars merge with smaller black stars they make ripples called gravity waves. This is just like the cycle of life that's mimicked all over ponds on Earth. As nature repeats, space is just a bigger bowl. Laws of constancy make everything spin. Constance is six energies spinning around four laws. Black stars have to obey these laws.

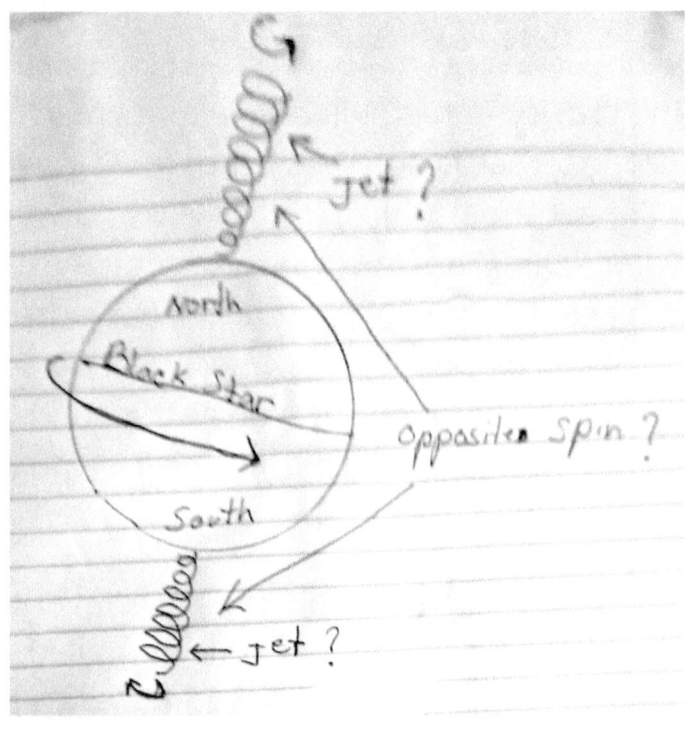

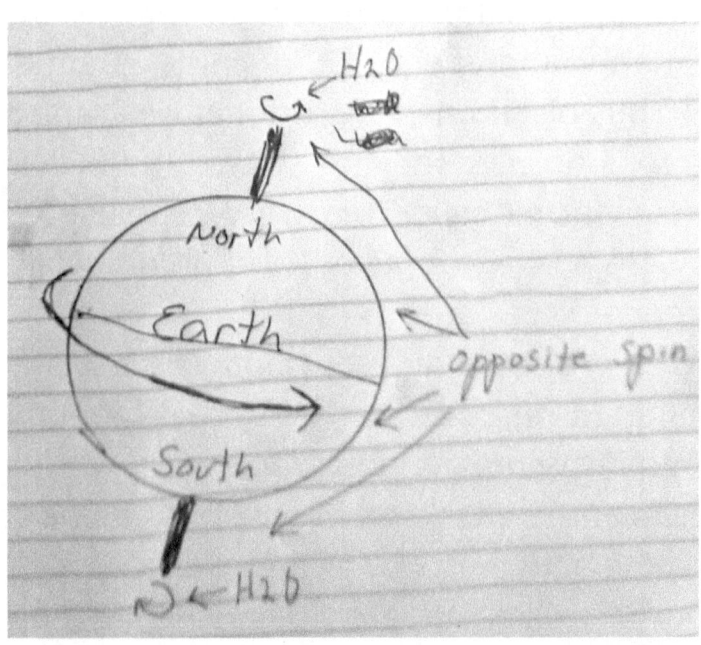

Before time = 0

Splitting gravity in half will divide and create a low pressure orb that is surrounded by high pressure. This will have a natural circling effect and will spin like water falling down a hole. Except the fourth room (space) has two holes that we call north and south. On Earth water spins one direction on the north side but the opposite way at the south pole. Black stars mimic this, and it's the last phase of recycling mass.

All energy plays a major role in space. Everything gets recycled or deposited in a bank called black stars. Constance is like understanding how energy works in a circular pattern. This gives a natural spin with matter and energy. Stars come in almost all colors, but black ones are the heaviest and most dangerous objects, moving around the Milky Way like magnetic pinballs. They will roll whichever way they were pushed or pulled, and nothing can stop it except a bigger one.

When one black star eats another, this is called combining, but the opposite of combining is dividing. This means that black stars will divide-opposite and then merge. Black stars can be stable or in motion. Here is why all interstellar space travel should be on high alert for these floating, opposite icebergs. Most astronomers will call a black star a black hole, but is it? This hole is just a gravity anomaly! Black stars are really white but will shine opposite or in reverse. Black stars always pull in; however, weak gravity can go both directions. It is unclear how big they can get, Black stars have no choice but to obey the laws of equals, opposites, actions, reactions, and spin.

DARK MATTER.

75

What is dark Matter or Dark energy?

Dark matter is a real freak of nature that may never get solved; however, there is a possibility it could be floating ghosts of negative energy. This idea comes from the eternal universe. The fifth room has blobs of positive energy everywhere, like an endless puzzle game, some can be larger or denser. This is the fuel score for a never-ending bomb. Some of the denser blobs will fall through and compact with some energy left. Like tiny pieces of bone after the fire is out, this leftover energy would be like floating ghosts of negatively charged dark matter.

Perhaps another possibility could come from north and south holes from the space bubble. These holes would have less resistance. Could this explain how space can continue to fill with matter as it expands? How can we equate the loss of energy, like those from gravity waves, and is this equal to the loss of mass like from black stars? If so, space could be working in a perfect circle. Its design is simple but incredibly accurate.

Could space be like a perfect light bulb? The mass from stars could represent the elements and negative energy is the oxygen, or the vacuum pressure, so when combined it radiates out a new energy called light; a perfect gift from energy itself, like a white rose exposing everything except dark matter!

EINSTEIN

Back when Einstein was dissecting relativity, he commented that romance was like the opposite. Perhaps when Einstein was young, he found himself hiding away in silence, trying to make sense of the obvious. Then, in his later years, he found himself hiding away from higher thinking-but confusing types of people that needed help keeping their feet on the ground.

In the world of atomic physics, Einstein would often get his feet wet, but he chose not to submerge too deep and referred to this science as spooky. Is this because atomics could be too dangerous? Einstein had many quotes, but the one I think most about is when he said that he had complete faith in space being simple but choosing not to reveal any more. I could not help myself to find this interesting but disturbing.

Einstein survived two world wars and understood more than just science, also understanding human nature. Here could be a true master Jedi—a man who could balance reality better than an average person.

Could Einstein have known that all energies are balanced between positive and negative charges? So if the fifth room is positive, and the fourth room is mostly negative- then opposites do attract. This is called expansion!

THE HUBBLE

The Hubble telescope is like an eye in the sky looking at time, frozen still by just flipping opposites, going from a telescope to a microscope to view the universe. The Hubble is like a big frozen eyeball floating in space, getting microscopic images of the smallest crystals of light. This universe has unfolded itself to incredible images of the past and can unveil our future as everything stays in constant motion in space. The Hubble has shown our closest neighbor, the Andromeda galaxy, to be a very ancient but hostile place with a high percentage of black stars. These are perhaps remnants from smaller galaxies in the past.

The Milky Way's fate will be no different, as our galaxy is but a drop of water in an ocean of unknown size and dimension. Could this fourth room be donut shaped with a massive void in its center? Just think of the Milky Way galaxy as the third room—a very extremely hostile place where radiation and gamma rays can penetrate flesh and bone like bullets. Our sun creates a solar wind that protects us from these bullets, but here in the first room, Earth is what protects us from these solar winds, and the moon helps shield us from rocks. From this viewpoint, step back and just look at the sun, Earth, and moon, and imagine that they're like our parents protecting us. The sun, moon and the stars were created in a divine way that reflects from a garden planet we've inherited. There are so many questions to ask without answers. The beginning of wisdom is to know that we don't know.

I would like to ask, but only crickets are listening and don't understand, and like us, we have become crickets in a garden that only birds and fish can hear.

And wait!

THE TREE OF POWER

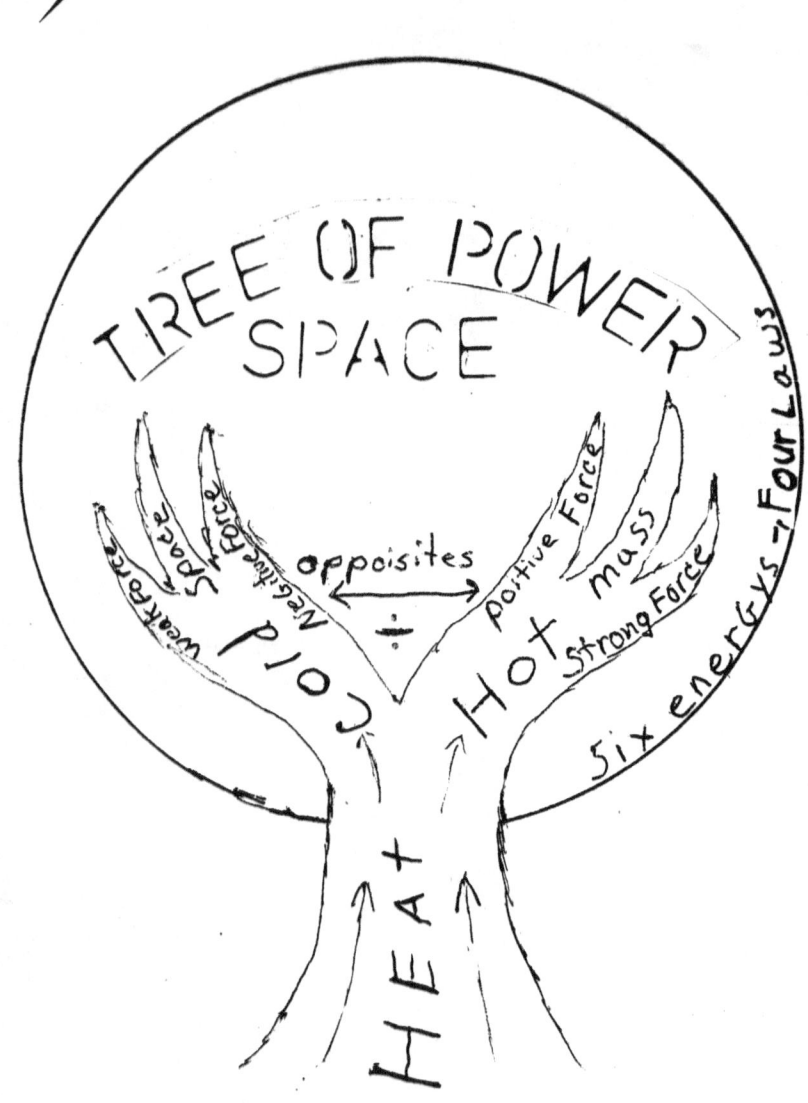

Look at this orb. It's an expansion of space. Inside the bubble, all energies doubles, but time and gravity have been slowed by 50 percent. To give, you have to take; this is like conduction. This is simple math. From the very beginning till now, all the energies in the new universe have been divided, starting with three energies in the old universe to six in the new universe. This all started with the making of an opposite universe by going from heat to cold; however, it takes up a lot of energy to make something new. This energy could have come from gravity and would have left a weaker gravity behind.

Negative energy was made opposite of positive energy. Gravity divided the new universe, setting up a chain reaction called equals and opposites, a simple universe. This is where the laws of motion spin like gears to slowly turn function and force order, called time. Regular gravity is all mighty and never-ending, but the gravity we feel is weaker. Gravity has formed a bubble. Weak gravity is what we feel now and is being pulled by the fifth room. This will force expansion. Here are two forces that will dictate a perfect circle.

The family tree of power can be viewed as simple math. If you double the laws from the fifth room to the fourth, then the power to multiply has to come from somewhere, and only strong gravity has the power to do this—but it comes with a price. Gravity has to divide, and dividing is the opposite of multiplying. The big bang is all about dividing energy, dividing energy 50/50 will force out two new laws called equals and opposites. Weak gravity is simply teetering itself outward and is being pulled by regular gravity like a inflatable balloon with no air source just exhaust from a never ending bomb,this exhaust is called negative-energy.

TREE HUGGER

If I was a tree, thin what could I do?

I'll be forced to grow fast and hide under the blue. Trees are like a coat that will cover the lands.That can help change the weather while catching a tan. These are my friends, this is no doubt, They ask nothing in return except the air we breathe out. The Forest is a perfect cathedral carved from nature that is under constant assault from overpopulation, greed and fossil fuels. Fortunately there are multitudes of good nature people around this beautiful blue planet that really do care about nature.

However could it be said that trees have betrayed all animals on earth by being loyal to us? It's taken millions of years for trees to cover the lands. When man discovered fire, we only needed a few thousand. At that time there was a peaceful balance between trees and nature. But within nature itself are basic habits, like when a wolf sees a rabbit it's instinct is to kill it! Even with a full stomach. So why are people this way with trees? Rabbits should be a sign of peace, I believe that trees are a sign of hope. Lightning fires once divided the forest from the fields, but now it's man. Always cutting and burning. Most trees like bears will hibernate in the winter and wake up to an uncertain arena. One can argue that a bark from a wolf can be part of its protection but not for trees. A bark from a tree cannot protect it from the saw. Most trees will stand tall with nowhere to run and hide from Violent Men with spinning swords of climate altering machines. These are beavers controlled by grizzly bears that come with gold and will always say "it's just not enough". Yes I'm a tree hugger, ironically being a carpenter I do try to preserve what was lost but what could truly be lost is a environmental conscience we once shared with the forest. Some people who love being in canopies share a connection with the animals of the forest, While others will mock you for being a monkey. That's okay, These people are like elephants with powerful legs that will topple and devour its own food sources. Naturally native animals will try to roam these ancient woodlands only to be replaced with barbed wire. Soon a trade dispute will occur that would allow Buffalo hunters access to lands that were once covered with beautiful trees. America was once a garden protected by natives who shared a cycle of life that ensured a continuous bounty without the plow. Most of the new people coming from the oceans arrived to escape prosecution from their previous homelands. Unfortunately they brought the seeds of diseases and iron that would cripple the stewards that protected the forest, while leaving only small reservations with no trees that survived extinction.

Now, with so many hungry elephants that will drink and be merry in large artificial arenas built from the bones that once protected the air we breathe. These proud people can be seen driving large climate controlled vehicles and will snicker at others that are less fortunate. The real unfortunate ones will be the next generation to inherit a planet that has lost most of its protection above and below the sea. So please join me and others as we chain ourselves to what trees are left because we can't breathe underwater to save the reefs.

WE THE PEOPLE

This eye is your conscience looking at two mirrors.

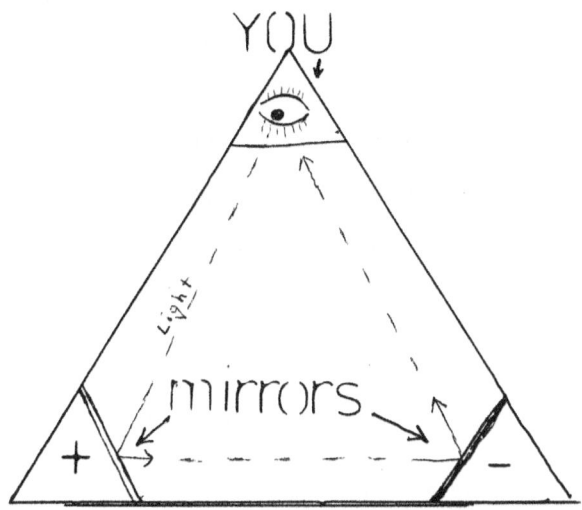

One mirror is backward, and flipping one energy or mirror backward or opposite creates choice. This eye is your conscience and will never close. We must sleep, but our conscience will still roam around!

Rest can prevent so many health issues, because a conscious reflection can be exhausting! Sleeping and drifting out of consciousness is like recharging your batteries for a new day with clean mirrors. Now you can blaze a trail and follow your own lead. always try to follow your good nature and pursue what seems right. Deceiving others will only lead to conflicts with yourself and will create your own weather for a bad environment. Wisdom can be our best ally.

Gaze ahead and ask questions that even you can't answer. Asking questions can be like planting seeds. These seeds could grow and find you, and your questions may be answered. Planting seeds could be like writing songs, these songs should be kept private. Others might mock what they don't understand. Humans can be like sheep grazing in a predictable way . But a true Jeti can see above the sheep. We can see the wolves counting sheeple.

Processed news and politics has ways to herd good people into pastures with no shepherd. Television, radio, and smartphones have become all-out assault weapons for elections. Here you will find aggressive people with loud voices that will expect loyalty for causes that will someday deplete itself. But for now we are the viewers that cover the fields. We are the wolves that pick off the weak. We are the heathens that kill the strong. Yes—we have become mostly the sheep that can only whisper, because our voices have been silenced by these howling wolves. If we could talk out loud we would say there aren't enough shepherds—and we may be right—but multitudes of people will still choose to hide behind loud excuses, while others will come asking to serve you and your country. Beware– some are wolves in disguise, as man's best friend, but look again. There are always nicely dressed, smooth-speaking men and women who are well funded; these are handpicked people of high IQ who lack common sense.

Expecting a higher income will lower a person's standards. Now these are new politicians (actors) that will eat gold between plays on stages built by oligarchs, spotlighted by pickle face executives and watched by the gullibles.

Cover your bondage boxes deep, and cloak yourself from this all-seeing eye that is watching you.

A true Jeti respects all brothers and sisters. We share a house with (Earth) and will never bow to a false leader.

So take another look at your phone,radio and television. could these be bondage boxes of too much negative energy? Maybe, but don't let your guard down, because positive energy can be just as deceiving. Here, power is for sale; however, some executives (who I call wolves) will expect loyalty, using the entertainment industry, combining fear and lobbyists to support their cheesy actors in power. Presiding like overpaid lawyers with no conscience, they will scramble our brains with truth mixed with lies.

We all know a child is taught to always look both ways before crossing the street, but who is teaching us? These information highways can easily flip to misinformation. Somehow many have forgotten how to look both ways before crossing the internet and flipping channels. In time we could be flipped to support our deceivers. Soo many viewers and surfers are getting suckered for smaller waves, with longer commercials. All by those actors we trusted. When we can't find them, we'll....it gets easy to forget and just move on to the next feel-good scandal. Try balancing common sense between good and evil while stepping backward for a better view. Television executives, for example, who wish to indulge in other people's choices will only thirst for a larger audience while disfiguring its viewers with inappropriate content.

These bondage boxes are like projecting crystal balls connected to countless mirrors. This is a spider web mostly controlled by oligarchs, hedge vultures, and gorilla executives standing on top of skyscrapers, always looking down on the subjects they feed on while ignoring massive amounts of climate pollution. A butterfly is free to vote but a spider will kill it. These boxes represent a lot of over- competitive entertainment meat markets that are stacked with wolves. They all share a ferocious appetite for money, fame and power. Here are factions that have loud speakers and spotlights that shine on stages built by control freaks.

So who protects our children from this? Your bondage boxes are competing for its future, not theirs. These reckless people are using programs and shows that teeter on the fence of being legal. This fence was built in the past to protect Americans from hearsay and the young from immorality. Now this fence has been trampled by overfed elephants. Deregulation changed averythang.

Science is a study of the repeats of the laws of nature that cannot be changed by humans. People can only change their own laws, and no other changes of human laws have had such a profound impact on its very own future than what took place in 1987 and beyond. At that time, the powers to the courts and their jesters seized an opportunity to change certain laws. These laws were only changed to serve leverage to the wolves that sharpen their fangs through a microphone and camera's eye. Timing couldn't have been better for the pride; more deregulation took place right before this new revolution of the internet, cable, and smartphones. These new laws were the beginning of hearsay, which was like a contagious virus in which foolish personal opinions with loud voices were and are being hurled into our living rooms. Dishonest political leaders at that time allowed deceiving news to flourish while setting in stone a new era of deceptive content for younger viewers. This is when Hollywood and the government became bed partners. Before 1987 we were used to hearing stories like bridges to nowhere or the Iran-contra scandals, but now this type of news is swept under the rug.

Before the September 11th attack, we all witnessed the first reversed election that somehow flipped, and from that confusion came two more Vietnam-like wars. Presented by four loudspeakers that supported two new warlords. Many of these jesters are still in power. Legislation seems to have become altered by loud and unproven opinions. This will only benefit the factories where so many Americans can become properly processed.

Gravity seems to have a collapsing effect with mass, just like fear. Fear can be a force as strong as gravity and holds the power to collapse a person's ability to comprehend the truth. Fear can reveal itself from both truth and lies, but lies travel much farther. Promoting lies takes loyalty and money. A weak mind could find strength in fear. These bondage boxes are good at creating fear. Try stepping back and gaze at these nicely dressed, beautiful, and well-speaking people, and take note. They're lying to you, and they're good at it. That is why they're there. Their voices are loud. These are master speakers with powerful amplifiers blasting out of bondage boxes from a fox that has made its way into the fields and is dividing the vast herds of people. This clever animal is always tethered to executive gorillas that can be appealing to watch but can't be trusted. A fox is related to the wolf, and **we** the people have become the prey.

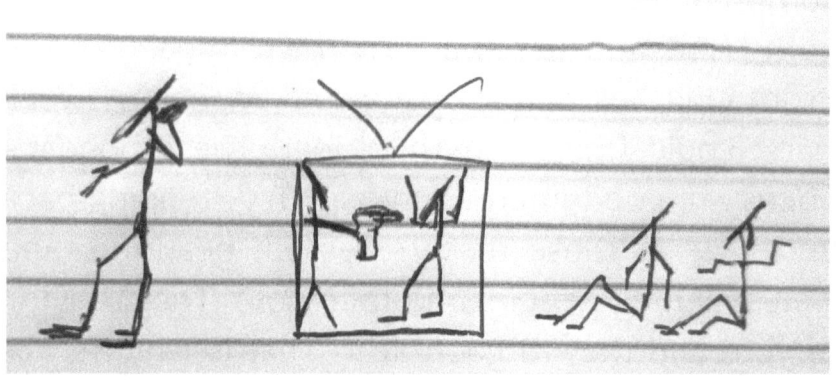

DOUBLE PYRAMID

Could Earth have been created to mimic the universe? Space can be viewed as a triple opposite. In space it is basically very hot or very cold. The three laws of hot and cold are the same: direction, charge, and temperature—a pyramid effect. But now look at Earth. Earth was divided perfectly between hot and cold, day and night, and charge (or north and south). Could it be possible to imagine Earth as a small pyramid that could have been created inside a bigger one?

Could the Earth be a Proving Ground ?

Now imagine sitting down in front of a mirror—but this time don't get up and ask who we are. In space, this fourth room is a massive arena where everything dwells, and it's mostly made up of hydrogen, just like us. And in these rooms are rocks that are like bone, so inside these rocks is iron, just like blood. Now look at space again. It's hydrogen, carbon, and iron controlled by electromagnetic force. Now look again at the mirror, but this time open your eyes and ask, Are we not the same?

Carl Sagan.... referred to us as "space dust." Earth is a tiny blue dot in the middle of a narrow beam of visible light. However I see earth as a tiny fish in a Cosmic bowl of depleted energy. Alpha Centauri is the closest neighbor to the sun and is a triple solar system with an unknown number of planets. Trying to find another Earth has a chance of one and sixty billion or hitting the jackpot on a slot machine with about fifty Wheels. The Earth and Moon are in perfect alignment with the Sun, so I would like to know what this means? In time this moon during an eclipse will start getting smaller, but why is it perfect now? what is going on! is this just another incredible coincidence? Asking questions that no one can answer can be unpopular. Earth could be considered a very incredible place, but it does have its dangers, the biggest being too much crude oil. Oil has led to a displacement of wealth and overpopulation. This will add conflict for depleting resources. In 2014 I visited the Philippines islands and witnessed what overcrowding can do to a small nation under strict laws that can incriminate population control. With that said I was surprised to see so many people in good spirits despite their harsh living conditions and lack of economic opportunity. This helped me look at a different side of humanity that we somehow tend to ignore. Here in these small islands are the multitudes of young families that struggle with poverty. Oceans can seem like prison walls. Passports are in demand but in short supply. I do believe larger nations that struggle with healthcare could benefit from the Filipino people. Pain from the 15th century once quoted that each government must keep up with current times or face history to repeat itself...

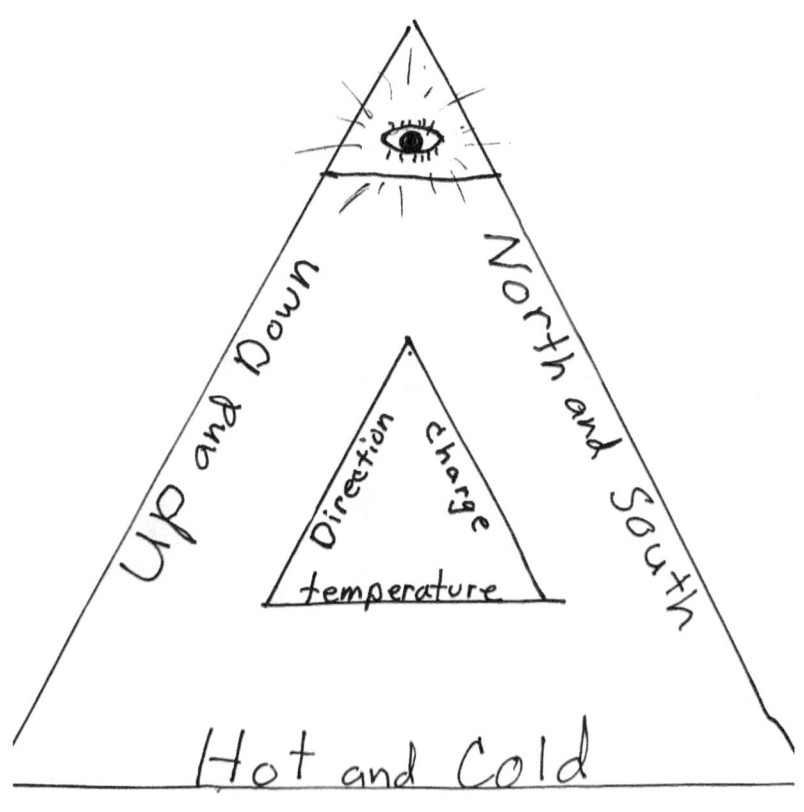

The Three Laws of Positive and Negative Energy

END GAME

FIFTH ROOM
(ETERNAL UNIVERSE)

Could this eternal room be packed with heat and strong gravity. This heat is a total repelling charge. In this place positive energy functions differently. In space, however, the fourth room's positive force has been fused with the negative force and will make gravity neutral, uniting into two new energies called the electromagnetic force which could actually be a combination of three energies. Now you have positive energy and negative energy. They are the same as hot and cold. Gravity divided and united all six energies to form space. The fifth room may have had only three energies that could have divided to create six energies to form the fourth room. The power to divide will come from gravity. The fourth room is expanding because of a never-ending bomb.

This never-ending bomb is believed to be the skin, or the outer layer, of the space bubble. This should be a massive temperature flip. As positive charge is ignited, it flips from super-hot to its opposite and falls into the fourth room as negative energy. Space is a ball of depleted energy, and with no energy, this will make a perfect highway for new energies to cross with almost no resistance.

What, if anything, was to cross into this fifth room? Everything would be black and crazy. Without negative energy (the electromagnetic glue), anything and everything will simply break apart.. Traveling through this place would be like following a circle and trying to find the end. If you were to go back into the fourth room from here you will be deep-fried and fall back into the space bubble as negative energy. The fourth and fifth rooms are opposites.

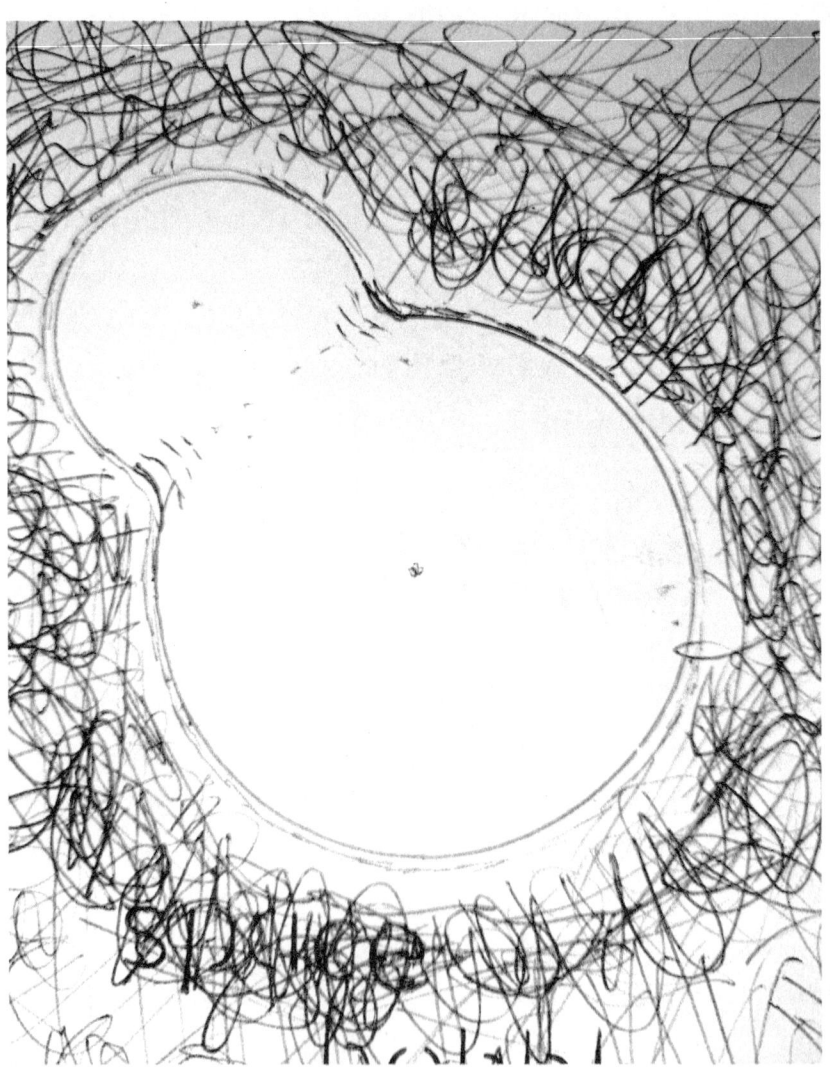

What would happen if our expanding universe or the space bubble was to encounter another bubble? Common sense would say that only cold bubbles can grow, but we just don't know. If a hot universe existed out there, then white holes could exist there too. This would mean that a hot universe could function in the opposite direction and might collapse.

This seems unlikely, but what if two growing space bubbles were to cross? How would two frozen fourth rooms in an endless ocean of chaos react? Maybe it would be just like two bubbles in a bathtub. Two will become one; two galaxies can cross through each other because of the massive amount of space in between but not space bubbles. Two bubbles that are fixing to touch will buckle from a lack of fuel that it needs to expand. These two bubbles will fuse together. The smaller one will be swallowed, sending out massive waves inside and expanding the bubble. Like a large whale eating a smaller one in an endless ocean of chaos, the bigger whale will become much larger. Combining these two whales together would make titanic waves instead of ripples. This could disrupt the spin of the fourth room and increase or decrease the pull of gravity. Disrupting gravity that is perfectly balanced can be extremely fragile and cave in. In the fourth room, balance is everything, everything is balanced between charges. The word equal really means balance, negative energy is balanced with positive energy, weak gravity is balanced with strong gravity and space is balanced with mass. The force has opened my eyes to possibilities that I could never have imagined before.

Have you ever wondered what's inside a black hole? Look around—you're in one.

Try understanding the laws of motion and forget about atomics. Atomics come last on the family tree of power. It's too confusing. Einstein called this science "spooky." Negative energy has flipped a gap between the laws of motion and quantum mechanics. Is it possible that most of the elements on the chart would have been created last in the vacuum of space? This would have been perhaps the first hundred million years after the big bang, making new types of matter in a primeval bowl of soup that is constantly mixing up new recipes for this expanding room. This new universe would expand out, while galaxies are drifting apart in clusters.

The fourth room is like a colossal cosmic clock with countless spinning gears. The Mayans were obsessed with trying to figure this out, and like us, they created several different clocks and calendars that only their nobles understood.

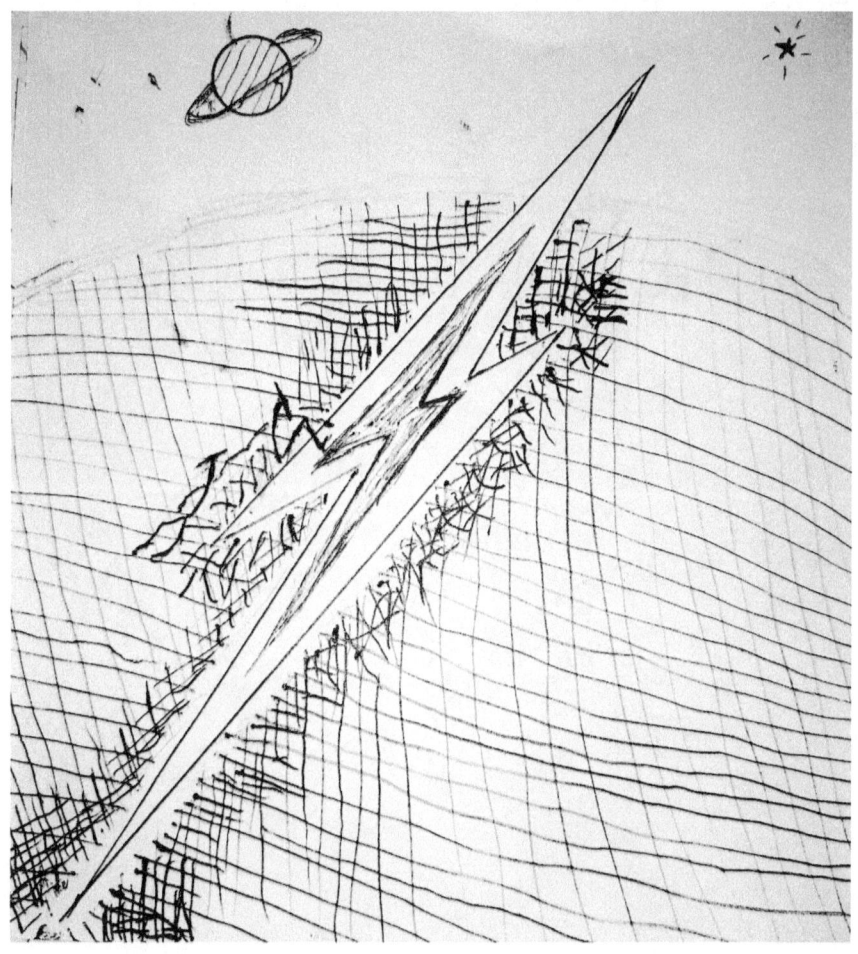

Today's high speed calendar and clock makers seem to be tripping over each other. Quantum confusion has a ripple effect that attracts a cornucopia of never-ending questions. This pattern is a slow process of elimination. Just like the medical world of uncovering disease, that means you jump from one breakthrough to the next, which takes time. However, time is always in demand, but for now it's in short supply.

THE END

What is the end? Could it have something to do with its opposite, like the beginning? Light is the beginning. Like morning, every day has to fall to night. Every beginning has an end. A cycle of life starting with equals and opposites. To understand the entire new universe as an orb, you have to go back to before it all got started.

You have to think of the entire old universe like a flat disc of just water that goes on forever and never stops in any direction. If you were to blast a hole in this flat disc of just water that represents gravity, because gravity is the only force here in this fifth room, the blast will keep going outward in all directions and never stop like a tidal wave. But this new hole has caved in, and then you've made a hole of low pressure with only one dimension. It's just a new hole.

Now think of this water as regular gravity that will fall into this hole as a weak spot or weak gravity. Water will fall in and start filling up this hole, creating a bigger hole, and on the edges of this hole is called expansion. It'll just keep eroding away like a waterfall effect as this hole gets bigger. Now regular gravity will fill this hole forever, creating the only black hole that can exist. This is called the big bang, and like time, this is the only beginning that has no end. This is a chronology of how everything could have taken place in the expanding universe from the very beginning to the present time. This is for a deeper understanding of how all energies work together and how they connect to us.

Please take consideration for our only home amongst extremely hostile surroundings (space).